Workplace Violence Prevention Strategies and Research Needs

Report from the Conference

Partnering in Workplace Violence Prevention: Translating Research to Practice
November 17–19, 2004, Baltimore, Maryland

DEPARTMENT OF HEALTH AND HUMAN SERVICES
Centers for Disease Control and Prevention
National Institute for Occupational Safety and Health

This document is in the public domain and may be freely copied or reprinted.

Disclaimer

Mention of a company or product does not constitute endorsement by the National Institute for Occupational Safety and Health (NIOSH). In addition, citations to Web sites external to NIOSH do not constitute NIOSH endorsement of the sponsoring organizations or products. Furthermore, NIOSH is not responsible for the content of these Web sites.

The recommendations and opinions expressed in this document are those of individual participants in the conference and may not represent the opinions and recommendations of all members, or of the organizations they represent.

Ordering Information

To receive documents or other information about occupational safety and health topics, contact NIOSH at

NIOSH Publications Dissemination
4676 Columbia Parkway
Cincinnati, OH 45226–1998

Telephone: **1–800–35–NIOSH** (1–800–356–4674)
Fax: 513–533–8573
E-mail: pubstaft@cdc.gov
or visit the NIOSH Web site at **www.cdc.gov/niosh**

DHHS (NIOSH) Publication No. 2006–144
September 2006

SAFER • HEALTHIER • PEOPLE™

Foreword

Since the 1980s, violence has been recognized as a leading cause of occupational mortality and morbidity. On average, 1.7 million workers are injured each year, and more than 800 die as a result of workplace violence (WPV) [Bureau of Justice Statistics 2001; BLS 2005]. These tragic deaths and injuries stress the need for a proactive and collaborative WPV prevention effort at the national level.

As part of its WPV Research and Prevention Initiative during 2003, the National Institute for Occupational Safety and Health (NIOSH) convened a series of stakeholder meetings that focused on various types of WPV and the industries and occupations at risk. For example, separate meetings addressed domestic violence in the workplace, violence in heath care facilities, violence in retail settings, and violence against law enforcement and security professionals. The purpose of these meetings was to bring together subject matter experts from business, academia, government, and labor organizations to discuss current progress, research gaps, and collaborative efforts in addressing WPV. One of the recurring discussion points that emerged from the meetings was the need for a national conference on WPV prevention.

In November 2004, NIOSH assembled a diverse group with representatives from various disciplines and organizations that have a stake in reducing the toll of WPV. This landmark conference—*Partnering in Workplace Violence Prevention: Translating Research to Practice*—was held in Baltimore, Maryland, on November 15–17, 2004. The sessions were structured to give participants an opportunity to discuss the current state of national research and prevention efforts. The intent was to draw out their best professional judgments on (1) identification and implementation of effective prevention programs and strategies, (2) identification of barriers to prevention and steps for overcoming them, (3) current research and communication needs, and (4) the advancement of research and prevention through effective partnerships.

This report summarizes discussions that took place during the conference. The report does *not* include a documented review of either the literature on WPV in general or intervention effectiveness research in particular. In addition, the authors have consciously avoided adding the NIOSH perspective to this report or otherwise augmenting its content. We have preferred to represent as accurately as possible the information, ideas, and professional judgments that emerged from the discussions that took place at the Baltimore workshop.

In my view, the November conference was very successful. This report provides the following:

1. A useful direction for overcoming current barriers and gaps that impede collaborative research, prevention, and communication work

2. An emerging collective vision (based on input from participants) of effective WPV prevention programs that employers and practitioners can consider now

3. A discussion of the research and partnerships needed to advance WPV prevention

I believe that this report will further raise awareness of this national problem and point the way to increased knowledge about the risks, causes, and prevention of WPV. In addition, this report will help companies initiate, improve, and evaluate their WPV prevention efforts. Ultimately, the document will help to accelerate the current downward trends in injuries and deaths from on-the-job assaults.

John Howard, M.D.
Director, National Institute for Occupational
 Safety and Health
Centers for Disease Control and Prevention

Contents

Foreword iii

Abbreviations vii

Acknowledgments viii

Conference Planning Committee Members ix

1 Introduction 1
    1.1 Scope of Workplace Violence (WPV) 1
    1.2 Background: surveillance, research, and prevention 3
    1.3 Methods and objectives 7

2 Barriers and Gaps that Impede WPV Prevention and Strategies to Overcome Them 8
    2.1 Barriers to WPV prevention practice 8
    2.2 Gaps in WPV prevention research 11

3 WPV Prevention Programs and Strategies 14
    3.1 Strategies or approaches that may apply to more than one type of WPV 14
    3.2 Strategies specific to Type I (criminal intent) prevention ... 16
    3.3 Strategies specific to Type II (customer/client violence) prevention ... 17
    3.4 Strategies specific to Type III (worker-on-worker) prevention ... 17
    3.5 Strategies specific to Type IV (personal relationship violence) prevention ... 18

4 Research Needs for WPV Prevention 19

5 Linking Research to Practice 21

6 Partners and their Roles 23
    6.1   NIOSH 23
    6.2   Other Federal partners 24
    6.3   State agencies 24
    6.4   Private-sector companies, corporations, and alliances 24
    6.5   Business and community organizations 24
    6.6   Insurers 25
    6.7   Law enforcement 25
    6.8   The legal profession 25
    6.9   Academic research institutions 25
    6.10 The media 25
    6.11 The medical community 25

	6.12 Worker assistance programs	25
	6.13 Social advocacy organizations	26
	6.14 Other national organizations	26
7	**Conclusions**	**27**

References 28

Appendix 29

Abbreviations

ASIS	American Society for Industrial Security
BLS	Bureau of Labor Statistics
CAL/OSHA	California/Occupational Safety and Health Administration
CFOI	Census of Fatal Occupational Injuries
IPV	intimate partner violence
MADD	Mothers Against Drunk Driving
NIOSH	National Institute for Occupational Safety and Health
NTOF	National Traumatic Occupational Fatalities
OSHA	Occupational Safety and Health Administration
PSA	public service announcement
WPV	workplace violence

Acknowledgments

The National Institute for Occupational Safety and Health (NIOSH) recognizes the cosponsoring organizations that provided financial support for various aspects of the conference: Corporate Alliance to End Partner Violence, Verizon Wireless, State Farm Insurance Company, American Society for Industrial Security (ASIS) International, American Association of Occupational Health Nurses, Liz Claiborne, the Occupational Safety and Health Administration (OSHA) and Injury Prevention Research Center—University of Iowa.

The conference and this summary report would not have been possible without the enthusiastic efforts of the conference planning committee (see the list that follows). Members worked diligently to structure the conference, and NIOSH thanks them for their time, energy, and insight. Corinne Peek-Asa, Jonathan Rosen, Kim Wells, and Carol Wilkinson presented plenary session overviews and facilitated breakout sessions on each type of workplace violence (WPV). Meg Boendier, Stephen Doherty, Kathleen McPhaul, and Corinne Peek-Asa provided the working group session summaries that formed the basis for this report. NIOSH appreciates the technical contributions of all these persons and their dedication to the understanding and prevention of WPV.

Nancy Stout, Director, NIOSH Division of Safety Research, and Tim Pizatella, Deputy Director, Division of Safety Research, provided guidance and support. Lynn Jenkins, NIOSH, prepared and presented a summary of the conference themes and issues for use in preparing this report. The following NIOSH staff members prepared, organized, and reviewed conference material: Lynn Jenkins, Matt Bowyer, Dan Hartley, Kristi Anderson, Barbara Phillips, and Brooke Doman. Matt Bowyer and Herb Linn summarized conference notes and summary reports, drafted the text, and revised the report.

Jane Weber and Gino Fazio provided editorial and production services.

Conference Planning Committee Members

Matt E. Bowyer, Chair
Division of Safety Research
National Institute for Occupational Safety and Health

Gregory T. Barber, Sr.
Occupational Safety and Health Administration
Directorate of Enforcement Programs

Patricia D. Biles
Workplace Violence Program Consultant

Bill Borwegen
Director
Occupational Health and Safety
Service Employees International Union

Ann Brockhaus
ORC Worldwide
Occupational Safety and Health

Pamela Cox
Division of Violence Prevention
National Center for Injury Prevention and Control

Butch de Castro
American Nurses Association
Center for Occupational Health and Safety

Stephen Doherty
Doherty Partners LLC

Mary Doyle
John Hopkins School of Public Health

Paula Grubb
National Institute for Occupational Safety and Health

Michael Hodgson
Veterans Affairs/ Veterans Health Administration

E. Lynn Jenkins
National Institute for Occupational Safety and Health

Kathleen McPhaul
American Association of Health Nurses
University of Maryland School of Nursing

Susan Melnicove
Director of Education
ASIS International

Corinne Peek-Asa
Associate Director
College of Public Health
University of Iowa

Robyn Robbins
Assistant Director
Occupational Safety and Health Office
United Food and Commercial Workers International Union

Rashuan Roberts
National Institute for Occupational Safety and Health

Gene Rugala
Supervisory Special Agent
National Center for Analysis of Violent Crime
Federal Bureau of Investigations

Robin Runge
Director
American Bar Association
Commission on Domestic Violence

Linda M. Tapp
Administrator of Consultants Practice Specialty
American Society of Safety Engineers

Mary Tyler
U.S. Office of Personnel Management

Kim Wells
Executive Director
Corporate Alliance to End Partner Violence

Introduction

In North Carolina, two masked men walked into a food mart, killed the 44-year-old male co-owner by shooting him several times with a handgun, ripped away the cash drawer, and fled from the scene.

In Massachusetts, a 27-year-old mechanic in an autobody shop was fatally shot in the chest by a customer after they argued about repairs.

In Virginia, an ongoing argument between two delivery truck loaders at a furniture company distribution warehouse ended abruptly as one pulled a gun and shot the other to death.

In South Carolina, a 24-year-old woman who worked in a grocery store was taken hostage at gunpoint and then shot to death with multiple shotgun blasts by her 20-year-old ex-boyfriend.

These tragic examples of violence in U.S. workplaces represent a small sample of the many violent assaults that occur in U.S. workplaces annually.*

1.1 Scope of Workplace Violence (WPV)

According to the Bureau of Justice Statistics, an estimated 1.7 million workers are injured each year during workplace assaults; in addition, violent workplace incidents account for 18% of all violent crime in the United States [Bureau of Justice Statistics 2001]. Liberty Mutual, in its annual *Workplace Safety Index*, cites "assaults and violent acts" as the 10th leading cause of nonfatal occupational injury in 2002, representing about 1% of all workplace injuries and a cost of $400 million [Liberty Mutual 2004]. During the 13-year period from 1992 to 2004, an average of 807 workplace homicides occurred annually in the United States,

---

*These fatal, gun-related cases do not represent the huge number of violent incidents that result in nonfatal injuries or no injuries, or that involve other types of weapons. Also, these cases do not adequately represent the many industry sectors and worker populations that face the risk of violent assault at work.

according to the Bureau of Labor Statistics (BLS) Census of Fatal Occupational Injuries (CFOI) [BLS 2005]. The number of deaths ranged from a high of 1,080 in 1994 to a low of 551 workplace homicides in 2004, the lowest number since CFOI began in 1992. Although the number of deaths increased slightly over the previous year in both 2000 (677) and 2003 (631), the overall trend shows a marked decline [BLS 2005]. From 1992 through 1998, homicides comprised the second leading cause of traumatic occupational injury death, behind motor-vehicle-related deaths. In 1999, the number of workplace homicides dropped below the number of occupational fall-related deaths, and remained the third leading cause through 2003. In 2004, homicides dropped below struck-by-object incidents to become the fourth leading cause of fatal workplace injury (see Figure 1) [BLS 2005].

It is not altogether clear what factors may have influenced the overall decreasing trend in occupational homicides for the period 1992 through 2004, and whether the decreasing numbers will be sustained in subsequent years. Since robbery-related violence results in a large proportion of occupational homicides, certain trends (e.g., economic fluctuations) are likely to have contributed to the decreasing toll. The reduction may partially stem from the efforts of researchers and practitioners to address

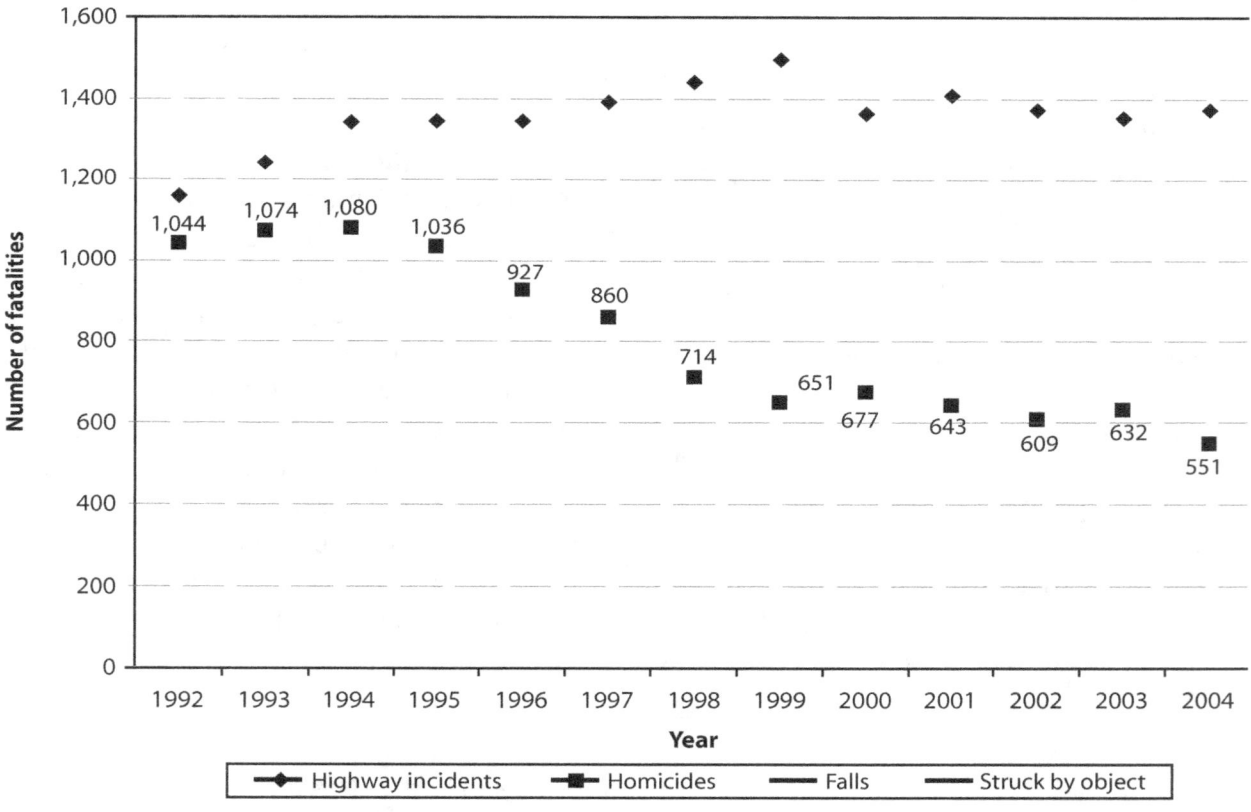

Figure 1. *The four most frequent fatal work-related events, 1992–2004.* NOTE: *Data from 2001 exclude fatalities resulting from the September 2001 terrorist attacks. (SOURCE: U.S. Department of Labor, Bureau of Labor Statistics, Census of Fatal Occupational injuries, 2004.)*

robbery-related WPV especially through intervention evaluation research and dissemination and implementation of evidence-based strategies. The reduction may be partially explained by the efforts of Federal, State, and local agencies and other policy-makers to develop statutes, administrative regulations, and/or technical information for WPV prevention as a result of improved recognition and understanding of the risks for WPV. Whatever the reasons behind the trend, future research and prevention efforts should focus on identifying, verifying, and replicating successes—such as reductions in robbery-related (Type I) violence—and identifying and addressing those types of WPV where little or no change has occurred. The fact that violence-related deaths increased over previous years' totals in both 2000 and 2003 raises questions about the sustainability of the overall downward trend and whether the occupational homicide experience in the United States may in fact be leveling.

A few of the violent incidents that occur in workplaces and result in deaths or serious injuries to workers are reported widely and prominently on TV and radio broadcasts, newspaper pages, and media Web sites. As indicated, WPV incidents arise out of a variety of circumstances: some involve criminals robbing taxicab drivers, convenience stores, or other retail operations; clients or patients attacking providers in health care or social service offices; disgruntled workers seeking revenge; or domestic abuse that spills over to the workplace (see Table 1). More recently, the threat of another form of criminal violence—terrorism—hangs over the nation's workplaces. Yet many employers, managers, and workers are not particularly aware that the potential for violence is a risk facing them in their own workplaces. The public is generally not aware of either the scope or the prevalent types of violence at work. In fact, it has been only within the last two decades that the problem of violent workplace behavior has come into focus—largely resulting from improvements in occupational safety and health surveillance—as a leading cause of workplace fatality and injury in many industry sectors in the United States.

1.2 Background: Surveillance, Research, and Prevention

When the National Traumatic Occupational Fatalities (NTOF) surveillance system was developed by the National Institute for Occupational Safety and Health (NIOSH) in the 1980s, an accurate count of workplace traumatic injury deaths in the United States was available for the first time [NIOSH 1989]. In 1988, NIOSH published its first article disseminating data on the magnitude of the national workplace homicide problem [Hales et al. 1988]. This article presented results indicating that worker against worker violence, which continues to be emphasized by the media, is only a small portion of the WPV that occurs daily in the United States.

The U.S. Department of Labor, through its Occupational Safety and Health Administration (OSHA) and the BLS, brought increased focus on occupational violence through compliance, surveillance, analysis, and information dissemination efforts. Although no specific Federal regulations then (or now) addressed WPV, OSHA began to cite employers where violent incidents occurred under the General Duty Clause [29 USC* 654 5(a)(1)], which requires employers to provide safe and healthful work environments for workers. OSHA also provided and disseminated, through reports and the OSHA Web site, violence prevention guidance for high risk sectors and populations such as health care, social services, late-night retail establishments, and taxi and livery drivers. The BLS has clarified the

---

*United States Code.

injury and fatality risks to workers from violent incidents through its nonfatal and fatal injury surveillance and special analyses of characteristics of occupational violence.

In the mid 1990s, as more researchers were becoming engaged in the study of occupational violence, the California Occupational Safety and Health Administration (Cal/OSHA) developed a model that described three distinct types of WPV based on the perpetrator's relationship to the victim(s) and/or the place of employment [Cal/OSHA 1995, Howard 1996]. Later, the Cal/OSHA typology was modified to break Type III into Type III and Type IV, creating the system that remains in wide use today [IPRC 2001]. (See Table 1.) This typology has proven useful not only in studying and communicating about WPV but also in developing prevention strategies. Certain occupations, such as taxicab drivers and convenience store clerks, face a higher risk of being murdered at work [IPRC 2001], while health care workers are more likely to become victims of nonfatal assaults [NIOSH 2002].

Since nearly all of the U.S. workforce (more than 140 million) can potentially be exposed to or affected by one of the four types of WPV, occupational safety and health practitioners

Table 1. Typology of workplace violence

Type		Description
I:	Criminal intent	The perpetrator has no legitimate relationship to the business or its employee, and is usually committing a crime in conjunction with the violence. These crimes can include robbery, shoplifting, trespassing, and terrorism. The vast majority of workplace homicides (85%) fall into this category.
II:	Customer/client	The perpetrator has a legitimate relationship with the business and becomes violent while being served by the business. This category includes customers, clients, patients, students, inmates, and any other group for which the business provides services. It is believed that a large portion of customer/client incidents occur in the health care industry, in settings such as nursing homes or psychiatric facilities; the victims are often patient caregivers. Police officers, prison staff, flight attendants, and teachers are some other examples of workers who may be exposed to this kind of WPV, which accounts for approximately 3% of all workplace homicides.
III:	Worker-on-worker	The perpetrator is an employee or past employee of the business who attacks or threatens another employee(s) or past employee(s) in the workplace. Worker-on-worker fatalities account for approximately 7% of all workplace homicides.
IV:	Personal relationship	The perpetrator usually does not have a relationship with the business but has a personal relationship with the intended victim. This category includes victims of domestic violence assaulted or threatened while at work, and accounts for about 5% of all workplace homicides.

Sources: CAL/OSHA 1995; Howard 1996; IPRC 2001.

and advocates should be concerned. Examples of high-risk industries include the retail trade industry, whose workers are most often affected by Type I (criminal intent violence), and the health care industry, whose workers may generally be affected most by Type II (client, customer, or patient violence). Although all four types of WPV can potentially occur in any workplace, Type III (worker-on-worker violence) and Type IV (personal relationship violence, also known as intimate partner violence), are more likely to occur across all industry sectors.

WPV includes a much wider range of behaviors than just overt physical assaults that result in injury or death. Thus, WPV has been defined as "violent acts, including physical assaults and threats of assault, directed toward persons at work or on duty" [NIOSH 1996]. It is widely agreed that violence at work is underreported, particularly since most violent or threatening behavior—including verbal violence (e.g., threats, verbal abuse, hostility, harassment) and other forms, such as stalking—may not be reported until it reaches the point of actual physical assault or other disruptive workplace behavior.

Most of the research that was conducted over the last half of the decade of the 90s was published in scientific and professional journal articles. Figure 2 shows the dramatic increase in the number of research articles published in the medical literature that dealt with WPV from the 1980s, when the occupational fatality surveillance data first showed that occupational homicide was the second leading cause of traumatic occupational death, through 2004 [National Library of Medicine 2005]. Similar results were obtained in searches of the occupational safety and health, business, and social science literature.

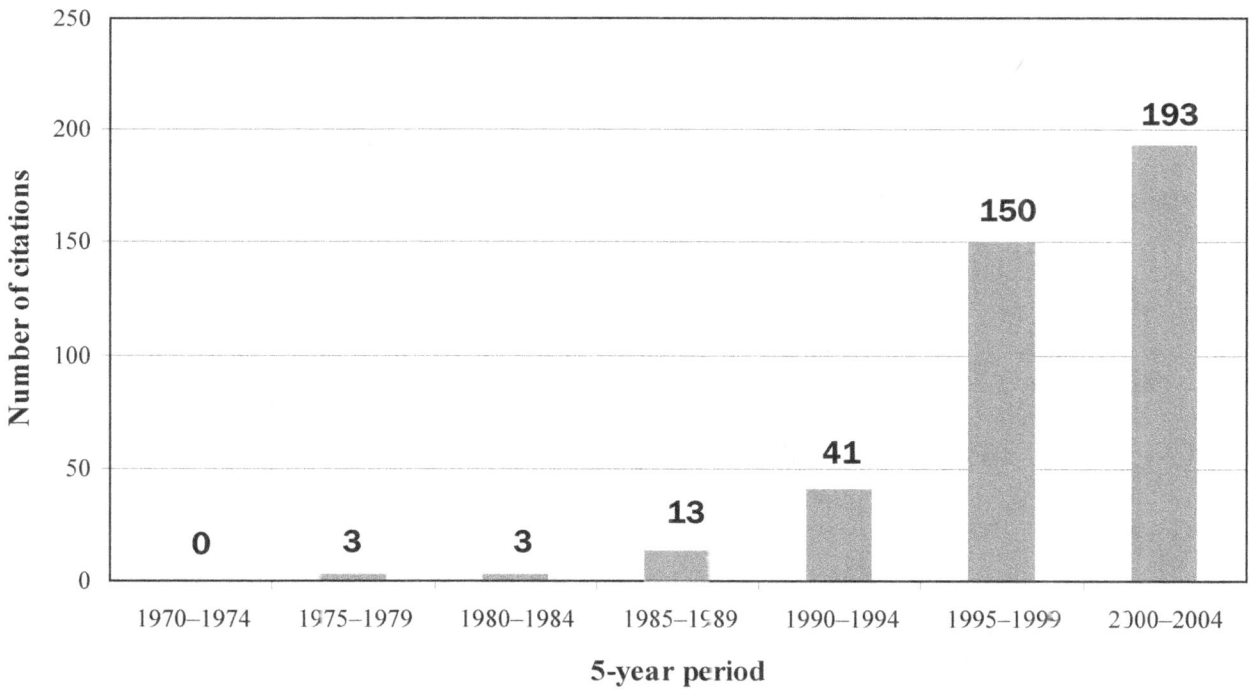

Figure 2. Medline entries for WPV for 5-year periods from 1970 to 2004.

In April 2000, the University of Iowa Injury Prevention Research Center sponsored a meeting entitled *Workplace Violence Intervention Research Workshop* in Washington, D.C. The workshop brought together invited participants to discuss WPV and recommend strategies for addressing this national problem. The workshop recommendations were published as *Workplace Violence: Report to the Nation* in February 2001 [IPRC 2001]. This report identified key research issues and called for funding to address these research needs.

In December 2000, Congress appropriated $2 million to NIOSH to develop a WPV Research and Prevention Initiative consisting of intramural and extramural research programs targeting all aspects of WPV. Most of the money was used to fund new research grants undertaken by extramural researchers. Intramural research efforts focused on collaborating with other agencies to collect improved data on WPV from workers and employers, convening a Federal interagency task force to coordinate Federal research activities, and collaborating with other groups to raise awareness of WPV and disseminate information developed through the Initiative.

In June 2002, the Federal Bureau of Investigation's National Center for the Analysis of Violent Crime hosted a symposium on WPV bringing together a multidisciplinary group to look at the latest thinking in prevention, intervention, threat assessment and management, crisis management, and critical incident response. The results were published in March 2004 in *Workplace Violence: Issues in Response* [Rugala and Isaacs 2004].

As part of the WPV Research and Prevention Initiative, NIOSH convened a series of stakeholder meetings on WPV during 2003. The purpose of these meetings was to allow subject matter experts from business, academia, government, and labor organizations to collectively discuss WPV in terms of current progress, research gaps, and potential collaborative efforts. Stakeholders with interest in the following topic areas met during the timeframes noted:

- **May 2003**—Violence in **Health Care** Settings
- **June 2003**—**Domestic Violence** in the Workplace
- **August 2003**—Violence in **Retail** Settings
- **November 2003**—Violence Against **Law Enforcement and Security** Professionals

One of the recurring themes that emerged from the stakeholder discussions was the need for a national conference on WPV prevention. In January 2004, NIOSH assembled a diverse planning committee to begin developing this forum. On November 15–17, 2004, NIOSH held, for the first time, a national conference on WPV prevention, entitled *Partnering in Workplace Violence Prevention: Translating Research to Practice* [NIOSH 2004].

This document is the final product resulting from the November 2004 conference. It summarizes what conference participants think are key strategies required for successful WPV prevention, further research and communication needs, barriers and gaps that impede prevention, and strategies for addressing them. The document also summarizes participants' thoughts about potential partners among Federal, State, and private agencies with the resources and skills necessary to collaborate in prevention efforts, conduct further research, and facilitate appropriate regulations. It is hoped that this report will serve several important purposes—to raise awareness of employers, workers, policy makers, and the public in general to the fact that WPV continues to be a major public health issue; to assist business and labor leaders in adopting effective prevention programs and strategies;

to aid researchers in identifying future projects; and to prompt government officials to consider more comprehensive national programs.

1.3 Methods and Objectives

During the conference, NIOSH assembled a diverse group of experts representing the four WPV typologies and the various disciplines engaged in WPV research and prevention efforts (see Appendix for a full list of participants). The conference was structured to give participants an opportunity to discuss successful WPV prevention strategies, barriers and challenges to WPV prevention, major research and information dissemination gaps, and potential roles for various organizations in WPV prevention over the next decade. In order to address the objectives in an effective manner, discussion points were posed to participants in breakout sessions that were divided into four WPV typologies: Criminal Intent (Type I), Customer/Client (Type II), Worker-on-Worker (Type III), and Personal Relationship (Type IV).

The objectives of the conference are reflected in the following instructions given to discussion participants:

- Identify successful WPV prevention strategies.
- Identify barriers and challenges to and strategies for implementing WPV prevention.
- Identify major research and information dissemination gaps in WPV prevention efforts.
- Identify existing and potential partners and their roles in advancing WPV prevention.

The conference included the following:

- State-of-the-art presentations from a panel of experts in each WPV type
- An evening group event featuring a series of one-act plays reflecting the human impact of violence in the workplace and cultural issues concerning violence
- Breakout sessions that addressed the four discussion points among each of the four WPV types
- Introductory and summary presentations of the discussions of each breakout session, by session moderators in plenary sessions
- A closing summary session

This report should provide a useful framework for thinking about the current state of WPV research, prevention, and communication activities in the United States. Chapter 2 presents a discussion of barriers and gaps that impede the development and implementation of WPV prevention programs. Chapter 3 summarizes the best WPV strategy/program practices presented by conference participants. This summary represents an implicit template for addressing WPV prevention on a company, corporate, agency, and national level and includes strategies both general and specific to the four types of WPV. Chapter 4 presents a discussion of general research needs; Chapter 5 addresses the importance of linking research findings to practical prevention efforts. One of two important themes of the conference—partnership—is the focus of discussion in Chapter 6. Included are some ideas about partners who should be involved in national, community, and company collaborations, and what they could be doing to address WPV. Chapter 7 provides some concluding thoughts and a call to action for potential collaborators in a national WPV prevention effort. The Appendix provides a full list of conference participants.

Barriers and Gaps that Impede WPV Prevention and Strategies to Overcome Them

Conference participants identified and discussed numerous barriers and gaps facing those working to implement existing strategies and programs addressing WPV prevention or those seeking to study and fill knowledge gaps related to WPV risks and prevention. In many cases, strategies for addressing and overcoming these barriers and gaps were proposed and discussed. Employers, managers and supervisors, safety practitioners, workers, members of the public safety and legal professions, researchers, designers and manufacturers of protective technologies and products, educators and communicators, and others—all face difficulties in the process of identifying, documenting, assessing, preventing, and communicating about violent workplace events.

This report essentially addresses two key audiences—those who are responsible for implementing WPV prevention programs in communities, companies, or workplaces (policy makers, employers, managers, safety and health practitioners, members of teams who come from multiple disciplines and perspectives, workers, etc.) and those who face challenges related to exploring and filling the gaps in our knowledge of WPV and WPV prevention (researchers). The most important barriers and gaps that impede the implementation of effective WPV prevention programs, strategies, and interventions usually depend on the particular organization in question, and sometimes the type of WPV. These issues are also discussed in Chapter 3.

Barriers impeding research efforts include lack of access to company and workplace information, and inadequate data to define the scope of WPV. Knowledge of intervention effectiveness is sparse, and information about the costs of both WPV incidents and prevention efforts versus benefits of specific prevention strategies and programs is lacking. Too little is known regarding specific characteristics of perpetrators, victims, companies, and circumstances surrounding violent events. These issues are discussed in Chapter 4.

2.1 Barriers to WPV Prevention Practice

2.1.1 Corporate Attitude, Denial

For some companies, a prevailing corporate attitude or denial of the potential for WPV, may be strong enough that employers and managers remain unconvinced that they need to address it. In some, violence is not recognized as a high priority among competing threats until a tragic, violent event occurs. In many organizations, the value of WPV prevention in reducing liability and turnover and increasing productivity is not well understood. Employers may also hesitate to explore WPV risks and issues because they are wary about negative company image, legal liability, assuming responsibility for workers' private lives, and worker enlightenment and empowerment. One line of thinking is that workers who become aware of these issues will certainly file complaints and claims. All of these factors are barriers to developing policies, providing training, recognizing and reporting violence, and developing and implementing WPV prevention programs. Workers readily perceive the lack of management acknowledgment of WPV and support for WPV prevention. On the other hand, corporate leaders who set out to raise awareness of WPV and improve workplace communication, demonstrate their acknowledgement of WPV and provide a foundation for improved reporting and risk assessment and program development and implementation.

2.1.2 The Culture of Violence; De-humanization of Workplaces

A profound barrier to WPV prevention is related to the culture of violence that permeates U.S. society, including workplaces.

2.1.3 Lack of Worker Empowerment

Violent events (especially Type 1 violence) are prevalent in small businesses where workers may lack a voice. Workers without a voice—that is, without a personal opportunity to provide their concerns or participate in leadership decisions—or without an advocate to speak for them, have great difficulty influencing the adoption or even the consideration of prevention programs. In many businesses, large and small, disconnects exist between management and workers that impede communication of concerns and collaboration.

2.1.4 Lack of Incentives, Disincentives

Conference participants believe that there are too few incentives for companies to implement WPV prevention programs. Few regulatory requirements address violence, many guidelines addressing violence are outdated, and the many legal issues prompted by Federal, State and local statutes, ordinances, and regulations present challenges to WPV prevention and can seem an impenetrable thicket. Current laws are often ineffective, unenforced, and inconsistent from State to State. Employers who might consider WPV prevention programs may feel at a competitive disadvantage if no mandatory, enforced regulations exist that cover the entire industry sector. If more compelling data on costs of violence and costs/benefits of prevention programs and strategies were available, companies would likely have more incentive to invest resources in WPV prevention programs. In addition, the positive effects of knowledgeable workers empowered to provide input and participate in planning and decision making, which can include improved safety and health, morale, efficiency, and productivity, provide an important incentive to management.

2.1.5 Lack of Awareness

For some, the most substantial barrier is simply a lack of awareness of the scope and importance of the problem on the part of employers and workers alike. This lack of awareness extends beyond company walls to all levels of the public and private sector and the general public.

2.1.6 Lack of Information, Access to Available Information

For other knowledgeable employers, a lack of access to risk information or evidence-based prevention programs or strategies may form a difficult barrier to action. Those programs and interventions that have been evaluated and shown effective in specific settings—for instance the interventions addressing violence resulting from convenience store robberies—have not been adopted in all workplaces where similar risks and circumstances are present. Further, they have not been evaluated for other workplaces and industry sectors facing similar risks. Many other programs and interventions that have been adopted or suggested for different types of WPV and different workplace settings and circumstances have not been rigorously evaluated, if evaluated at all. If evidence-based prevention programs and strategies are available, the information mostly resides in academia or government agencies. Researchers in academia and government are often satisfied with publication of their findings in the peer-reviewed literature, or lack the knowledge and means to further disseminate or translate their results for use in at-risk companies. As a result, employers may not be fully cognizant of the risks they and their workers face. Or, an employer or practitioner who is aware of the risks and who has the desire to establish and implement a prevention program may not be able to find or access evidence-based programs and interventions to use or choose from.

Among companies with WPV programs, some are reluctant to share WPV information (e.g., statistics, program information, effectiveness data), even among other departments in the same company. Privacy issues and proprietary and competitive attitudes may influence companies and agencies to guard their data, thus hindering data sharing. Compounding the effect of this barrier, researchers may fall short of the efforts needed to engage and partner with employers. This in turn limits the ability of researchers to determine characteristics of violent events, characteristics of those who are involved in and affected by them, and potential preventive approaches and their effectiveness. OSHA has guidelines for late night retail [OSHA 2004], but companies not under OSHA jurisdiction may not be aware of this information. Potential sources of information useful to businesses include police department crime prevention units, Web-based violence prevention and security sites, and insurance companies.

2.1.7 Lack of Communication/ Training

A major barrier to awareness and prevention of WPV is an overall lack of adequate and effective communication and training about what constitutes violence (definition); when violence has occurred (incident reporting); what the company does about violence (policy, procedures, disposition); and what peers and partners have learned and are doing (research, prevention, collaboration). In the pursuit of individual responsibilities and tasks, the importance of communication may be overlooked entirely or given a low priority among competing demands.

2.1.8 Lack of Resources

Many of the companies facing high risks of WPV are small companies with limited resources for research, prevention, and evaluation. In an increasingly pressurized economy and in the absence of sufficient cost-benefit data, prevention may be seen as an unwarranted expenditure rather than an investment with a return. Employers may address competing demands first unless a tragic violent event has already occurred to gain their attention and prompt action. Small companies often have neither the resources nor the staffs to address problems from a multi-disciplinary perspective.

2.1.9 Lack of Reporting

Violent events, wherever they occur, may not be reported for various reasons. When WPV occurs in companies that lack an enlightened, prevention-oriented culture, victimized workers may be inhibited from reporting single incidents or patterns of abusive behavior that would be reported and addressed in other companies. In such companies, victims or witnesses of violence may feel that nothing will be done if they do report. Otherwise well meaning employers or managers in companies that do not communicate to workers the behaviors that are considered to be violent, the mechanisms for reporting them, and assurances of security, confidentiality, and prompt response, may be unwittingly fostering a violent work environment that could ultimately experience a tragic, violent event. Too often, in the aftermath of such a tragedy, people remember precursor events or behaviors that should have prompted reporting, response, and intervention at the time they occurred. Sadly, failures to report verbal or physical abuse represent lost opportunities for prevention. Lack of reporting is also a fundamental barrier to effective surveillance, a critical component of WPV prevention at all levels, from company-level to national-level prevention.

2.1.10 Lack of Effective Followup to Reported WPV Events

Victims and recipients of threats or harassment expect a firm response. When management fails to respond promptly and firmly to reported WPV incidents, or does not follow through according to company policies and procedures, workers will perceive the lack of management commitment. Workers will then be hesitant to report future violent events and behaviors.

2.1.11 Lack of Written WPV Policy, Definitions, and Consequences (See Chapter 3.)

A company or corporation without a written WPV prevention program or policy may fail to provide critical information necessary to protect workers. Prevention efforts may not succeed without written documentation that includes company policy on WPV, definitions that clearly indicate what specific behaviors constitute WPV and are therefore prohibited actions, the specific consequences of those actions, who is accountable for the program and specific elements, and the roles and responsibilities of all workers.

2.1.12 Lack of Teamwork, Partnerships

Interdisciplinary and interdepartmental work is very difficult to initiate and maintain, even within the walls of one company. Effective programs require the combined efforts of employers, workers, law enforcement, and, for larger companies, the multiple departments with a stake in violence prevention and worker safety and health.

2.2 Gaps in WPV Prevention Research

2.2.1 Lack of WPV Intervention Evaluation Research

The ideal situation is for employers and practitioners planning and implementing WPV prevention programs to have credible, evidence-based interventions, strategies, curricula, and programs available. A primary research need in WPV prevention is to obtain evaluation data on strategies and interventions for a variety of workplace applications.

2.2.2 Lack of Best Practices for Implementation

The need for practical and proven guidance for program implementation goes hand-in-hand with the need for evidence-based prevention programs and strategies. Critical information about best practices for WPV programs is needed by employers and practitioners.

2.2.3 Lack of (or Inadequacy of) Data

Currently available data—based largely on police responses, emergency room admissions, workers' compensation claims, insurance payments to victims, and death certificates—do not reflect the scope of WPV, especially considering the incidence of noninjury and nonphysical events (e.g., threats, bullying, harassment, stalking). Reluctance on the part of corporations and companies to release data and to admit researchers into their environments for the purpose of collecting incidence data or evaluating interventions and programs impedes description of the WPV experience, as well as further investigations of causation and prevention. In addition, the victims and witnesses of WPV may be reluctant to report incidents for a variety of reasons. (See Section 2.1.9.) Aside from cultural and behavioral impediments to the acquisition of better data, technical issues exist that must be overcome. For example, a commonly accepted, operational definition of what constitutes WPV, while not perfectly fitting every scenario imaginable, will be necessary to the uniform collection of data. Standardized data collection using common definitions is essential to draw reasonable conclusions on effective prevention. Standardization may require the following:

- Better categorization of data
- Addition of key pieces of data to existing data sets
- Researcher access to data from companies and insurers, as well as workplaces

2.2.4 Lack of Information about the Costs of WPV; the Cost-Effectiveness of Prevention

The economics of WPV represents a substantial gap in knowledge. Understandably, employers desire and respond to solid, empirical cost data on actual and potential losses from WPV and benefits of prevention programs and interventions. They are interested in understanding costs relative to benefits and return on investment when it comes to development and implementation of programs. Employers may not expect each and every intervention to pay for itself, but they do seek a general idea of what to expect as a result of investing in prevention. A difficult concept to calculate and convey is the cost of a non-event—that is, one that is prevented through programmatic investment. Other important cost considerations include the loss of experienced workers and the resultant new personnel hiring and training costs.

2.2.5 Research and Communication Needs Specific to Type I (Criminal Intent) Prevention

Research is needed to provide evidence about effectiveness of specific environmental, behavioral, and administrative interventions in non-convenience-store settings. Also uncertainties about effectiveness of other suggested interventions require additional research to enable the attainment of consensus in controversial topics such as effectiveness of on-site guards, bullet-resistant barriers, certain training elements, and multiple clerks.

2.2.6 Research and Communication Needs Specific to Type II (Client on Worker Violence) Prevention

Currently, not enough is known about what produces violence in social service, health care, and other settings for worker-client interaction. What is known has not always been widely reported in the scientific literature or by the national media. Risk estimates are not available that clarify the influence of various situational and environmental factors.

2.2.7 Research and Communication Needs Specific to Type III (Worker on Worker) Prevention

Type III WPV is somewhat unique among the types in that most of the losses incurred as a result of a violent incident (e.g., losses related to the victim, the perpetrator, the damages, the productivity, etc.) are usually borne solely by one employer. More solid information about the direct relationship between the availability of reliable data and the opportunity for prevention, and the resultant potential for controlling costs through intervention, may be effective in persuading employers to share information and provide access needed by researchers.

2.2.8 Research and Communication Needs Specific to Type IV (Interpersonal Violence) Prevention

More rigorous, science-based efforts are needed in characterizing risk factors, costs, and effectiveness of WPV prevention programs and strategies addressing Type IV violence.

2.2.9 Other General Research Needs

Conference participants also offered a substantial list of research gaps, most of which were not discussed in detail.

According to Conference participants, research is needed to better understand the following:

- Variations in what is being done in individual businesses, industry sectors, law enforcement, and State and local governments
- What motivates businesses to take action in addressing WPV
- What types of regulation are effective
- Work organization and how it affects WPV prevention program implementation and impact
- Characteristics of both perpetrators and victims of each type of WPV
- Successful management systems for tracking WPV and followup activities
- What makes training effective—that is, what content, teaching methods, intervals, etc.
- How to disseminate information about effective violence prevention strategies and programs more widely and/or more appropriately
- How to effectively communicate
    — What WPV is
    — Protection and prevention as positive issues
    — The importance of scientific research in addressing WPV

WPV Prevention Programs and Strategies

This chapter is tailored for use by employers, managers, and safety and health practitioners who desire to develop and implement or evaluate company WPV prevention programs. Conference participants were asked to identify and discuss WPV prevention strategies, which may range from comprehensive, overarching company policies and programs to individual intervention strategies that seek to modify environment or behavior. Prevention programs and strategies that might offer increased protection against WPV in general are discussed first, followed by program and strategy elements that are unique to specific WPV typologies.

3.1 Strategies or Approaches That May Apply to More Than One Type of WPV

3.1.1 Management and Worker Commitment

The importance of management commitment to WPV prevention policies and programs cannot be overemphasized. Top management support helps ensure that adequate resources (including staffing) will be applied to the program, that the program will be launched from the top down, and that the effort will likely be accepted throughout the organization and sustained. Worker participation in planning, development, and implementation of programs and strategies is also important. The concept of dynamic commitment (i.e., involving both management and workers) in WPV prevention was discussed as a fundamental necessity underlying the allocation of adequate prevention program resources and the development of a violence prevention culture within an organization.

3.1.2 Multidisciplinary Team Approach to WPV Prevention

Another common theme voiced often during the conference was the need for collaboration of people from different disciplines, company units or departments, and levels of the organization. The involvement of persons with diverse expertise and experience is especially critical due to the depth and complexity of WPV prevention. Such teamwork is crucial for planning, developing, and implementing programs, as well as serving discrete functions, such as threat assessment teams formed to review and respond to reported physical, verbal, or threatened violence. Some of the key levels, disciplines, and departments mentioned included management, union, human resources, safety and health, security, medical/psychology, legal, communications, and worker assistance.

The pre-arranged use of outside expertise and collaboration with local law enforcement and local service providers was also offered as a way for companies to ensure effective programs, particularly in smaller companies with fewer workers, departments, and resources. Proactive planning/collaboration with local law enforcement may be helpful should an incident requiring police response occur.

3.1.3 Written WPV Policy/Program Tailored to Organization's Needs

A documented company policy/program must include definitions that clearly indicate what behaviors constitute WPV, including threatening or abusive physical and verbal behavior. Prohibited actions must be specified, and the specific consequences of those actions spelled out. A review and response system for all reported violent incidents must be in place, along with guidelines to assist those with the responsibility to review and respond. Specific procedures are needed for reviewing each reported incident, and mechanisms are needed to support and protect all affected persons. Ineffective followup undermines worker perception of management commitment and negates incentives to report incidents. Victims and recipients of threats or harassment expect a firm response. Review and response to reported violence might best be accomplished via a team approach (e.g., a threat assessment team).

Clear, precise definitions; mandatory comprehensive (all incidents) reporting; a structure and process in place for reporting; and timely and reliable review and response will all contribute to accurate reporting, which in turn enables precise risk assessment and dedication of appropriate resources to the program. These elements will also provide a basis for program evaluation. Programs that discourage reporting or blame the victim will not likely be successful At a minimum, the WPV policy/program should be reviewed annually but optimally can be easily tweaked as necessary. Good communication, confidentiality, teamwork, and accountability are musts. Communication must flow vertically (management to staff and vice-versa) and horizontally (i.e., across organizational divisions or departments). Communication can take many forms, and organizations should think outside the box when communicating information about WPV policies/programs. For example, information about company policy/programs can be communicated as inserts with pay stubs or on stickers for telephones. A WPV prevention program should be well integrated with other company programs.

3.1.4 Training

Training for both managers and workers is a key element in any WPV prevention program. The presence of management at training sessions can increase the visibility of the organization's top-level commitment to prevention. Training content may differ by type of WPV (see Sections 3.2 through 3.5), but in general, training (initially and on a recurring basis) should be provided on the hazards found in the organization's workplaces and in the organization's prevention policies and procedures, with emphasis on reporting requirements and the companies' review, response, and evaluation procedures. Training can be implemented from the top down, with managers and supervisors trained first. A train-the-trainer approach can be used, with supervisors responsible for training and evaluating training for their own staffs. Specialized training on creating a positive work environment and developing effective teams could be useful, as well as training to improve awareness of cultural differences (diversity) and to enable the development of workers' cultural competence.

3.1.5 Culture Change

Employers should examine the workplace to determine if there are cultural barriers to WPV awareness and prevention. If needed, the workplace culture should be modified to foster increased awareness of WPV and prevention, the clarification and enumeration of acceptable and unacceptable behavior, WPV reporting, availability of support for victims, and availability of help for perpetrators (if employed by the company, as in Type III and sometimes Type IV WPV).

3.1.6 Evaluation

Prevention programs and strategies should be evidence-based to the extent that evidence is available. However, often action must be taken before data can be collected and evidence of effectiveness obtained. It is crucial that companies make the effort to evaluate programs and strategies and cooperate with researchers in intervention effectiveness evaluation research. Employers may waste valuable resources on hazard control and training if evaluation procedures are not integrated into programs to measure impact. Information about successful programs and strategies must be effectively shared and communicated within companies and industry sectors and, where applicable, across sectors. While it is true that rigorous evaluation is challenging and often involves substantial cost, employers and researchers may, through collaboration, find ways to leverage their combined resources to selectively assess strategy and program effectiveness. In addition, such partnerships may provide a vehicle for sharing evaluation methods and results across many companies in an industry sector.

3.2 Strategies Specific to Type I (Criminal Intent) Prevention

The potential for Type I WPV exists across all industries but is prevalent in certain industries characterized by interaction with the public, the handling of cash, etc. Certain industries in the retail trade sector (convenience and liquor stores, for example) face higher than average risks. Specific environmental, behavioral, and administrative strategies have been implemented and evaluated as a result, particularly in convenience stores. A core group of interventions has been determined to be effective in convenience stores [Hendricks et al. 1999, Loomis et al. 2002], including the following:

1. *Environmental interventions*
   — Cash control
   — Lighting control (indoor and outdoor)
   — Entry and exit control
   — Surveillance (e.g., mirrors and cameras, particularly closed-circuit cameras)
   — Signage

2. *Behavioral interventions*
   — Training on appropriate robbery response
   — Training on use of safety equipment
   — Training on dealing with aggressive, drunk, or otherwise problem persons

3. *Administrative interventions*
   — Hours of operation
   — Precautions during opening and closing
   — Good relationship with police

— Implementing safety and security policies for all workers

Some interventions for convenience stores and other workplaces are controversial or not universally agreed upon by researchers. These instructions will require additional study, including the following:

- Having multiple clerks on duty
- Using taxicab partitions
- Having security guards present
- Providing bullet-resistant barriers

3.3 Strategies Specific to Type II (Customer/Client Violence) Prevention

3.3.1 Adequate Staffing, Skill Mix

One strategy that emerged from discussions of the Type II panel is that of ensuring adequate staffing and mix of skills to effectively serve client, customer, or patient needs. Low responsiveness and quality of service, which can result from inadequate staffing and skills of personnel, can produce frustration and agitation in clients or patients. For clients and patients, acute needs and accompanying real or perceived urgency combined with a history of violence, can place both staff and other clients/patients at risk. In addition, social services or health care workers who work alone may be vulnerable to assault, especially in worker-client relationships where the client has a criminal background or is mentally ill or emotionally disturbed.

3.3.2 Training

In addition to general training on WPV hazards and organizational policies and procedures, training specific to Type II violence could include recognition of behavioral cues preceding violence, violence de-escalation techniques and other related interpersonal and communication skills, new requirements (in health care) for patient seclusion and restraint, and proper restraint and take-down techniques.

3.3.3 Accreditation Criteria Tied to WPV Prevention

Another strategy would have accreditation bodies specify WPV program and training requirements as criteria for successfully meeting accreditation standards for social service and health care organizations and facilities. Specific programming and training in response to the demands of meeting such criteria should improve workplace protection from client/patient-based violence.

3.4 Strategies Specific to Type III Violence (Worker-on-Worker) Prevention

3.4.1 Evaluating Prospective Workers

Preventing worker-on-worker violence begins during the hiring process by employers who ensure that job applicants are properly and thoroughly evaluated by means of background checks and reference verification.

3.4.2 Training in Policies/Reporting

A key in worker-on-worker violence prevention is the comprehensive reporting of all prohibited behaviors among workers, including threatening, harassing, bullying, stalking, etc. Therefore, training during new worker orientation and subsequent refresher training should focus on company WPV definitions, policies, and procedures. Also, reporting should be strongly encouraged and supported.

3.4.3 Focus on Observable Behaviors

The perpetrators are present or former workers who usually have substantial knowledge of coworkers, physical surroundings, and often security and violence prevention measures. Observation and reporting of changes in behavior that become a concern are critical. Therefore, a successful prevention strategy will provide procedures for reporting and addressing observable behaviors that elevate to concerns. A strong company focus and emphasis on the observation and reporting of behaviors that generate concern, coupled with timely and consistent response (see Section 3.1.3), may help create a climate that deters violent behavior.

3.5 Strategies Specific to Type IV (Personal Relationship Violence) Prevention

3.5.1 Training in Policies and Reporting

To prevent Type IV violence, company policies and procedures must provide workers with clear-cut information about the nature of personal relationship or intimate partner violence (IPV), its observable traits and cues, and methods for discerning it in coworkers. Employers must train workers in what to do if they should suspect that a coworker is involved in interpersonal violence, either as a victim or perpetrator. Training should emphasize the relevant company policies and procedures.

3.5.2 A Culture of Support

A company should strive to create a culture of support for victims that includes assurances no penalties exist for coming forward, complete confidentiality will be observed, safety and security protocols will be implemented, and referrals to appropriate community services will be provided as options to workers. A company should also inform all workers about the consequences of being a perpetrator of IPV or any other form of WPV. The company should communicate clearly through policies and training that IPV behavior is inappropriate and will be dealt with. Furthermore, the company should attempt to create a culture that both supports victims and enables perpetrators to seek help. Providing referrals to appropriate community services and implementing long-term programs that address battering and bullying behavior are reasonable approaches.

Research Needs for WPV Prevention

This chapter presents WPV research needs, as identified by conference participants. It is tailored for use by researchers and research agencies and institutes engaged in, or interested in the study of WPV risk and prevention. Conference participants were asked to identify and discuss research and information dissemination gaps and offer strategies for filling those gaps. The overarching research needs identified by participants are presented below. It is hoped that this chapter can be used to inform the development of WPV research strategies and agendas. Further, it should be useful as a basis for formulating new research projects and for forging partnerships.

- **Establish national strategy/agenda.** Under the leadership of NIOSH, researchers from government, academic and private research institutes, businesses and associations, worker advocacy groups and unions, and other organizations, should collaborate with business leaders, safety and health practitioners and advocates, and other interested stakeholders to establish a national research agenda for WPV.

- **Conduct evaluation research.** A critical endeavor for research-business collaboration is the evaluation of prevention strategies and programs. The need is broad, spanning the wide range of prevention options, the types of violence, and the variety of industry sectors and individual workplaces. Evaluation research is also expensive and time consuming. Therefore, a strategic approach is needed in which priorities are carefully considered, costs are shared and resources leveraged, and results are widely disseminated especially to at-risk employers and workers and the associations and unions that represent them.

- **Develop consistent WPV definitions.** Employers, workers, and everyone else with a stake in occupational violence must have a clear, shared conception of what constitutes WPV. In addition to a shared conceptual definition, a consistent operational definition is needed for comparability in reporting and data collection.

- **Ensure consistent and universal reporting.** Reporting is an issue at the company level, at the industry level, and at the national level. Accurate and consistent reporting will enable both targeting of prevention research and assessment of trends and effectiveness.

- **Share data among partners**. Both businesses and agencies possess data on reported WPV incidents, which if collected, combined, and analyzed, would shed light on the broader WPV experience in the United States, and could potentially enable more focused and thereby cost-efficient prevention efforts in companies or sectors.

- **Conduct economics research**. Decision makers in the private sector are accustomed to analyzing costs, benefits, return on investments—in short, examining the bottom line issues that impact their businesses. Realistic assessments of the costs of WPV to businesses and society in general, and the cost-benefit of prevention, including cost-effectiveness comparisons of effective, focused prevention options are needed.

Linking Research to Practice

Research that has been conducted to date must be translated into practical preventive workplace action. It is clear that Conference participants see a gap in the availability of evidence-based prevention options for industry—that is, between what is known and what is applied in the workplace. As additional evaluation studies and demonstration projects are concluded, research findings of effective preventive interventions must be proactively translated into prevention products and technologies and transferred to and implemented in workplaces. The translation, transfer, and wider implementation of prevention strategies and programs may be as or more time consuming, costly, and challenging as their initial development and validation. However, the substantial input provided by participants in the conference suggests that an excellent opportunity exists for a broad, collaborative effort to do the following:

- Take stock of the knowledge base for WPV prevention.
- Explore the gaps in that knowledge.
- Prioritize needed research and information efforts.
- Identify opportunities for wider implementation of known effective prevention measures throughout workplaces, companies, and industries at risk.
- Identify and use existing data, findings, and knowledge that have yet to be translated and transferred to practical prevention technologies, products, interventions, strategies, programs, curricula, and recommendations.
- Collaborate and cooperate fully with potential partners to plan new research with implications for practical prevention.
- To help ensure such research, engage partners (particularly business and industry partners) earlier in the process of identifying problem areas and conceptualizing research projects and approaches.

Conference participants identified the following overarching needs in linking research to practice:

- Establish and maintain a clearinghouse of WPV-related information, particularly evidence-based programs and strategies.

  As in many domains, the volume of information related to WPV risks and prevention is growing. A daunting challenge looms in the organization, validation (assessment of reliability), tailoring, and distribution of information about WPV risks, prevention strategies and options, research findings, cost data, and other pertinent knowledge components. A key design objective should be easy access for employers and all other partners.

- Sponsor national, public information/education campaigns to raise awareness of WPV, emphasize the importance of prevention programs, and provide contact information for support services.

Wider awareness of the prevalence of WPV is needed among at-risk employers and workers, policy makers, media, and the general public. Federal government partners should help communicate existing knowledge, including what constitutes WPV, the types of WPV, the sectors and occupations at risk, and the critical roles of research, evaluation, and company policies and programs in the prevention effort. Information about availability of support services for organizations and individuals should be included. Such information might be particularly useful to companies seeking to develop and implement WPV programs, and individuals seeking help who may be either victims or perpetrators of WPV.

Partners and Their Roles

Participants in conference discussions repeatedly emphasized the importance of collaborating and partnering in WPV prevention—from the interdisciplinary and interdepartmental collaboration (so crucial to developing and implementing prevention programs) to national interorganizational partnerships (essential for advancing WPV research, implementing findings, and evaluating efforts). Partners need to be identified and engaged, roles need to be determined; agendas, strategies, and plans need to be developed; and programs need to be established, implemented, and evaluated.

This section identifies some of the partners (or types of partners) that participants suggested were necessary to the WPV research and prevention effort, as well as some of the roles and responsibilities that participants thought fit well with each based on their missions and activities.

6.1 NIOSH

NIOSH was recognized as a key organization, both in assuming specific roles and responsibilities suggested during the discussions and in facilitating the collective efforts of a wide range of partners. NIOSH was recognized for its current roles and activities as a leading research center, as a voice for objectivity in research and dissemination, as a strong advocate for identifying and improving effective research and prevention approaches, and as an organization that leverages resources, engages stakeholders, and prepares and disseminates information for the business community.

In addition to the NIOSH role in conducting, collaborating in, and coordinating WPV research, the following principal roles were suggested for NIOSH:

- Developing and keeping a clearinghouse of information about violent workplace events, model programs, data collection instruments, implementation practices, and other pertinent information potentially useful to employers and other stakeholders

- Developing (1) data-gathering standards for compiling data from disparate sources and (2) a reporting system that captures all WPV events—verbal abuse and other threatening behaviors as well as injury outcomes

- Leading an effort to make the issue of WPV more visible (through public information and education campaigns, for example)

6.2 Other Federal Partners

Suggested roles for other relevant Federal partners (such as OSHA, BLS, the Department of Justice, the National Center for Injury Prevention and Control, the Veteran's Administration, and other agencies that collect relevant data or regulate industry) in collaboration with NIOSH include the following:

- Coordinating the national WPV prevention effort over the next decade

- Forging a common definition with employer alliances and worker advocacy groups to identify the range of behaviors that constitute WPV

- Gathering data on the Federal workforce (the Nation's largest worker group)

- Implementing WPV prevention programs in Federal workplaces

- Ensuring and maintaining up-to-date statistics on WPV

- Adopting a partnership model to develop regulations addressing WPV

6.3 State Agencies

These roles were suggested for State agencies:

- Collaborating with Federal partners to embrace common definition(s) of WPV

- Quantifying victimizations among State workers and thereby adding to the available data

- Determining specific and relevant strategies for prevention in State government

6.4 Private-Sector Companies, Corporations, and Alliances

Roles suggested for private-sector companies, corporations, and alliances are the following:

- Contributing to the effort to forge common WPV definitions along with government agencies and worker advocacy groups

- Sharing data on WPV events as well as successes, problems, and methods to overcome barriers in implementing WPV prevention programs and strategies.

- Adopting WPV prevention strategies that have been recommended and verified by Federal agencies

6.5 Business and Community Organizations

Suggested roles for business and community organizations are as follows:

- Serving as conveners, bringing together factions of the community to engage in dialog, striving to comprehend the issue, and forging a coordinated response to WPV prevention

- Sharing prevention programs and strategies: a businesses-helping-businesses approach

- Assisting government, media, and educational institutions in increasing public awareness of WPV risks and prevention

6.6 Insurers

The following roles were suggested for insurers:

- Providing incentives, primarily by reducing workers' compensation premiums for employers who implement WPV prevention programs that demonstrably lower workers' compensation costs

- Supporting research that seeks economic evidence that violence prevention provides a return on investment to employers or other entities investing in WPV prevention

6.7 Law Enforcement

Roles suggested for law enforcement agencies include the following:

- Collecting more detailed data and standardizing definitions

- Disseminating evidence-based prevention information

- Providing assistance to businesses in taking prevention steps

- Participating in research efforts to address the prevention of workplace crime and violence

- Focusing on community-oriented policing

6.8 The Legal Profession

These roles were suggested for the legal profession:

- Appropriately balancing the need for collecting accurate WPV victimization data with the tangle of overlapping privacy interest laws

- Securing exemptions or waivers from existing privacy restraints in order to collect data

- Training attorneys to be sensitive and provide outreach to affected clients

6.9 Academic Research Institutions

The following roles were suggested for academic research institutions:

- Training new researchers entering the field

- Raising the research bar by setting the example in research and crafting violence prevention strategies based on findings

- Playing a proactive role in accessing private industry data

- Emphasizing in its law, business, and management curricula the dynamics of WPV and its impact on workers, families, and corporate health

6.10 The Media

The role suggested for the media was providing public service announcements (PSAs) in support of public information campaigns.

6.11 The Medical Community

The medical community's suggested role was to improve recognition and reporting of potential cases of injury or stress from WPV.

6.12 Worker Assistance Programs

Suggested roles for worker assistance programs were the following:

- Improving screening and recognition of potential WPV issues

- Being involved in response to WPV incidents to serve victim, witness, and co-worker needs

6.13 Social Advocacy Organizations

Roles suggested for social advocacy organizations were the following:

- Contributing to the effort to forge common WPV definitions with Federal, State, business, and labor partners

- Developing media campaigns following the model provided by Mothers Against Drunk Driving (MADD)

6.14 Other National Organizations

The following roles were suggested for other national organizations:

- Having safety and security specialists and organizations interact with research and regulatory communities to enable research-to-practice linkage (incorporate findings in their programs and procedures) and to provide expert input to researchers and regulators

- Having academic schools of architecture, urban planning, and civil engineering to interact with violence prevention partners to provide expert input to research and regulatory efforts and to incorporate safety and security considerations into their designs

Conclusions

This report summarizes the discussions that occurred during the conference *Partnering in Workplace Violence Prevention: Translating Research to Practice* in Baltimore, Maryland, in November 2004. Many ideas are presented about what is missing from the national effort to study and prevent WPV. Some gaps could be addressed by increasing intervention evaluation research; improving reporting, data collection, and data sharing; facilitating and enabling organizations to foster the dynamic commitment and cooperation of employers and workers; analyzing costs and cost-benefits; and improving organization and delivery of risk and prevention information. Other gaps are more specific to the types of violence, the various roles and relations among partnering organizations, or the industries and occupations involved.

Great strides have been made over the past two decades. Likewise, opportunities exist to address the barriers and gaps outlined in this report and to achieve a more coordinated, efficient, and cost-effective national effort to understand, control, and prevent violent incidents at work. These violent incidents damage or destroy the victims' sense of security, dignity, and (too often) their well-being and their lives. They represent a large toll to our society.

The key to the utility and impact of a report such as this is the extent to which people and organizations can visualize and initiate the efforts and partnerships needed to understand and reduce the risks of WPV within their spheres of influence. We encourage your interest, involvement, and collaboration in this effort.

References

BLS [2005]. Census of fatal occupational injuries. Washington, DC: U.S. Department of Labor, Bureau of Labor Statistics [http://www.bls.gov/news.release/pdf/cfoi.pdf].

Bureau of Justice Statistics [2001]. Violence in the workplace, 1993–1999: special report from the National Crime Victimization Survey. Washington, DC: U.S. Department of Justice, Bureau of Justice Statistics [www.ojp.usdoj.gov/bjs/pub/pdf/vw99.pdf].

Cal/OSHA [1995]. Cal/OSHA guidelines for workplace security. Sacramento, CA: California Occupational Safety and Health Administration [www.dir.ca.gov/dosh/dosh%SFpublications/worksecurity.html].

Hales T, Seligman PJ, Newman SC, Timbrook CL [1988]. Occupational injuries due to violence. J Occup Med 30(6):483–487.

Hendricks SA, Landsittel DP, Amandus HE, Malcan J, Bell J [1999]. A matched case-control study of convenience store robbery risk factors. J Occup Environ Med 41(11):995–1004.

Howard J [1996]. State and local regulatory approaches to preventing WPV. Occup Med: State of the Art Reviews 11:2.

IPRC [2001]. WPV: a report to the nation. Iowa City, IA: University of Iowa, Injury Prevention Research Center, February.

Liberty Mutual [2004]. Liberty mutual workplace safety index: the direct costs and leading causes of workplace injuries. Boston, MA: Liberty Mutual, 4 pp. [http://www.libertymutual.com/omapps/ContentServer?cid=1078439448036&pagename=ResearchCenter%2FDocument%2FShowDoc&c=Document].

Loomis D, Marshall SW, Wolf SH, Runyan CW, Butts JD [2002]. Effectiveness of measures for prevention of workplace homicide. JAMA 287(8):1011–1017.

National Library of Medicine. PubMed bibliographic database search conducted on-line (search strategy: "WPV" OR "occupational violence" OR "workplace assault" OR "occupational assault" OR "workplace homicide" OR "occupational homicide") [www.ncbi.nlm.nih.gov/entrez/query.fcgi?db=PubMed]. Date accessed: July 8, 2005.

NIOSH [1996]. Current Intelligence Bulletin 57: Violence in the workplace; risk factors and prevention strategies. Cincinnati, OH: U.S. Department of Health and Human Services, Centers for Disease Control and Prevention, National Institute for Occupational Safety and Health, DHHS (NIOSH) Publication No. 96–100.

NIOSH [1989]. National traumatic occupational fatalities, 1980–1985. Cincinnati, OH: U.S. Department of Health and Human Services, Centers for Disease and Control Prevention, National Institute for Occupational Safety and Health, DHHS (NIOSH) Publication No. 89–116.

NIOSH [2002]. Violence: occupational hazards in hospitals. Cincinnati, OH: U.S. Department of Health and Human Services, Centers for Disease and Control Prevention, National Institute for Occupational Safety and Health, DHHS (NIOSH) Publication No. 2002–101.

NIOSH [2004]. Partnering in workplace violence prevention: translating research to practice. Conference held in Baltimore, Maryland, November 15–17. Cincinnati, OH: U.S. Department of Health and Human Services, Centers for Disease and Control Prevention, National Institute for Occupational Safety and Health [www.cdc.gov/niosh/conferences/work-violence/].

OSHA [2004]. Guidelines for preventing WPV for health care and social service workers, OSHA 3148-01R. Washington, DC: Occupational Safety and Health Administration [www.osha.gov/Publications/osha3148.pdf].

Rugala EA, Isaacs AR [2004]. Workplace violence: issues in response. Quantico, VA: FBI Academy, Federal Bureau of Investigation, National Center for the Analysis of Violent Crime, Critical Incident Response Group [www.fbi.gov/publications/violence.pdf].

USC. United States Code. Washington, DC: U.S. Government Printing Office.

Appendix

Conference Participants

Participants in **Partnering in Workplace Violence Prevention: Translating Research to Practice**, Baltimore, Maryland, November 15–17, 2004

Jorge Acuna
Crisis Prevention Institute
jacuna@crisisprevention.com
262–317–3445

Steve Albrecht
Albrecht Training & Development
drsteve@drstevealbrecht.com
619–445–4735

Sania Amr
University of Maryland, School of Medicine
samr@epi.umaryland.edu
410–706–1466

Kristi Anderson
National Institute for Occupational Safety and Health
kanderson2@cdc.gov
304–285–6362

Debra Anderson
University of Kentucky
debra.anderson@uky.edu
859–257–3410

Susie Ashby
U.S. Army
Susie.Ashby@usag.apg.army.mil
410–306–1057

Jim Azekely
International Taxi Drivers Safety Council
Jim4Safety@aol.com
304–525–0902

Karen Baker
National Sexual Violence Resource Center
kbaker@nsvrc.org
717–909–0710, Ext. 101

Charlene Baker
Centers for Disease Control and Prevention
cbaker@cdc.gov
770–488–1737

Marianne Balin
Blue Shield of California Foundation
marianne.balin@blueshieldcafoundation.org
415–229–5215

Angela Banks
Centers for Disease Control and Prevention
abanks@cdc.gov
770–488–4273

Gregory T. Barber Sr.
Occupational Safety and Health Administration
Barber.Greg@dol.gov
202–693–2473

Cathleen Batton
Baltimore County Police Department
CBatton@co.ba.md.us
410–931–2145

DeAnna Beckman
University of Cincinnati
deanna.beckman@uc.edu
513–558–3951

Javier Berezdivin
AngerExperts.com
javierb@aol.com
305–757–2161

David Bernstein
Forensic Consultants, PC
drbernstein@forensicconsultants.com
860–833–7732

Jean Berrier
Convenience Store Safety Committee
Davidberrier@cox.com
757–425–2396

Patricia Biles
Biles and Associates
bandbiles@aol.com
202–484–8226

Vicki Black
U.S. Army
Vicki.Black@usag.apg.army.mil
410–278–7155

Heather Blackiston
DAOHN
heather.blackiston@mbna.com
302–458–0586

Barbara Blakeney
American Nurses Association
Bblakene@ana.org

James Blando
New Jersey Department of Health
james.blando@doh.state.nj.us
609–777–3039

Chris Blodgett
Washington State University
blodgett@wsu.edu
509–358–7679

Meg Boendier
Workplace Violence Advocate
mboendier@aol.com

Matt Bowyer
National Institute for Occupational Safety and Health
mbowyer@cdc.gov
304–285–5991

Mark Braverman
Marsh USA
mark.braverman@marsh.com
202–263–7938

Ann Brockhaus
ORC Worldwide
brockhaus@orc-dc.com
202–293–2980

Jean–Pierre Brun
Laval University
jean–pierre.brun@mng.ulaval.ca
418–656–2405

Tanya Burrwell
American Psychological Association
tburrwell@apa.org
202–336–6049

John Byrnes
AON Aggression Management
DrJohnByrnes@AggressionManagement.com
407–804–2434

Nancy Carothers
Convenience Store Safety Committee
nmcaroth@pacbell.net
650–867–8349

Norman Caulfield
Commission for Labor Cooperation
ncaulfield@naalc.org
202–464–1107

Rebecca Cline
Ohio Domestic Violence Network
rclineodvn@aol.com
330–725–8405

James Coleman
National Council of Chain Restaurants
jcoleman@constangy.com
571–522–6100

Joanne Colucci
American Express Company
joanne.colucci@aexp.com
877–208–7492

Phaedra Corso
Centers for Disease Control and Prevention
pcorso@cdc.gov
770-488-1734

Francis D'Addario
Starbucks Coffee Company
fdaddari@starbucks.com
206-318-8736

Linda Dahlberg
Centers for Disease Control and Prevention
ldahlberg@cdc.gov
770-488-4496

Lucia Davis-Raiford
Miami-Dade County Government
davisra@miamidade.gov
305-375-2685

David Dawson
Empowerlink Threat Management
empowerlink@iprimus.ca
905-945-0101

Butch de Castro
American Nurses Association
Bdecastro@ana.org

Katherine Deitcher
DAOHN
katherine.deitcher@mbna.com
302-432-0025

Richard V. Denenberg
Workplace Solutions, Inc.
worksolutions@taconic.net
518-398-5111

Tia Schneider Denenberg
Workplace Solutions, Inc.
worksolutions@taconic.net
518-398-5111

Frank Denny
Department of Veterans Affairs
frank.denny@mail.va.gov
202-273-9743

Bob DeSiervo
American Society of Safety Engineers
bdesiervo@asse.org
847-768-3402

Carmen Dieguez
Miami-Dade County Government
ccd@miamidade.gov
305-375-2682

Edward DiSabatino
MBNA Bank
francis.dixon@mbna.com
302-457-2167

Francis Dixon
MBNA Bank
francis.dixon@mbna.com
302-457-2179

Stephen Doherty
Doherty Partners LLC
s.doherty@dohertypartners.com
617-393-9928

Brooke Doman
National Institute for Occupational Safety and
 Health
bdoman@cdc.gov

Robert Dorsey
Santa Clara County Domestic Violence
 Council
bdorsey@cisco.com
408-525-0107

Terrie Dort
National Council of Chain Restaurants
dortt@nrf.com
202-626-8183

Rosanne Dufour
Commission des normes du travail
rosanne.dufour@cnt.gouv.qc.ca
418–380–8521

Carole Dupere
Commission des normes du travail
carole.dupere@cnt.gouv.qc.ca
514–873–4947

Jennifer Edens
Federal Bureau of Prisons
jedens@bop.gov
202–514–4492

Gerhard Eisele
Oak Ridge Associated Universities
eiseleg@orau.gov
865–576–2208

Rosemary Erickson
Athena Research Corporation
rjerickson@athenaresearch.com
605–275–6028

Shelley Erickson
U.S. Department of Agriculture, Food Safety
    and Inspection Service
shelley.erickson@fsis.usda.gov
515–727–8981

Don Faggiani
Police Executive Research Forum
dfaggiani@policeforum.org
202–321–9354

Gary Farkas
Psychologist/HR Consultant
gary@garyfarkas.com
808–521–2433

Richard Fazzio
Occupational Safety and Health
    Administration
fazzio.richard@dol.gov
617–565–8110

Giuseppe Fichera
Department Occupational Health Clinica Del
    Lavoro Luigi Devoto
giuseppe.fichera@unipd.it
0039/02/5454091

Dawn Fischer
U.S. Army
Dawn.Fischer@usag.apg.army.mil
410–278–3609

John Flood
U.S. Air Force
jbflood68@yahoo.com
405–234–2262

Pamela Foreman
University of California San Francisco
pforeman@itsa.ucsf.edu
707–292–1886

James Fox
Northeastern University
j.fox@neu.edu
617–373–3296

Kelley Frampton
U.S. Bureau of Labor
frampton_K@bls.gov
202–691–6189

Roland (Ron) Fravel III
U.S. Department of Agriculture, Food Safety
    and Inspection Service
roland.fravel@fsis.usda.gov
202–690–1999

Eric Frazer
Forensic Consultants, PC
Yale University School of Medicine
drfrazer@forensicconsultants.com
203–624–0111

Tom Galassi
Occupational Safety and Health
    Administration
Galassi.Thomas@dol.gov
202–693–2100

Linda Garber
State Farm Insurance
linda.c.garber.cyp2@statefarm.com
434–872–5958

Chantenia Gay
U.S. Department of Labor
gay.chantenia@dol.gov
202–693–4906

Dorothy Goff
Consultant
651–777–0311

Teague Griffith
National WPV Prevention Partnership/SCDVC
scdvcmail@domesticviolence.net
509–487–6783

Paula Grubb
National Institute for Occupational Safety and Health
pgrubb@cdc.gov
513–533–8179

Jeffrey Hagen
The Community College of Baltimore County
jhagen@ccbcmd.edu
410–780–6955

Mary Jane Haggitt
University of Washington
rredcar@comcast.net
770–712–8113

Patrick Hancock
Baltimore County Public Schools
phancock@bcps.org
410–887–4133

Sarah Hansel
VA Maryland Health Care System
sarah.hansel@med.va.gov
410–642–2411, Ext. 5499

Jim Hardeman
WPV Prevention, Inc.
JAAMES73@AOL.COM
508–746–6021

Randy Harper
HCR Manor Care
rharper@hcrmanorcare.com
410–480–2333

Daniel Hartley
National Institute for Occupational Safety and Health
dhartley@cdc.gov
304–285–5812

Gail Heller
gheller@choicesdvcols.org
614–258–6080

Jennifer Hilliard
American Association of Homes and Services for the Aging
jhilliard@aahsa.org
202–508–9444

Michael Hodgson
Veterans Health Administration
muh7@mail.va.gov
202–273–8353

John Howard
Director, National Institute for Occupational Safety and Health
jhoward1@cdc.gov
202–401–6997

Terri Howard
Target Corporation
terri.howard@target.com
612–761–4214

Lee Husting
National Institute for Occupational Safety and Health
ehusting@cdc.gov
404–498–2506

Lisalyn R. Jacobs
Legal Momentum
ljacobs@legalmomentum.org
202–326–0040

Lynn Jenkins
National Institute for Occupational Safety and Health
ljenkins@cdc.gov
304–285–5822

Barbara Kabrick
International Taxi Drivers Safety Council
barbj2799@comcast.net
509–475–3842

Ann Kaminstein
DV Initiative
ann@dvinitiative.com
617–306–6969

Michelle Keeney
U.S. Secret Service
Michelle.Keeney@usss.dhs.gov
202–406–5205

Gwendolyn Keita
American Psychological Association
gkeita@apa.org
202–336–6044

Susan Kindred
CCBC-Catonsville
skindred@ccbcmd.edu
410–455–5133

Trina King
U.S. Postal Service
trina.l.king@usps.gov
202–268–3981

Carrie Kirasic
Weber Aircraft
Clkirasic@msn.com

Nicholas Lamis III
Miami–Dade County Government
NLamis@miamidade.gov
305–375–2680

John Lane
The Omega Threat Management Group, Inc.
OMEGATMG@AOL.COM
310–551–2063

Douglas Leach
Blue Shield of CA Foundation
douglas.leach@blueshieldcafoundation.org
415–229–5462

Rocky Leavitt
Ken Bu Kai, Inc., Martial Arts
rocky@kenbukan.org
270–982–3548

Theresa Leavitt
Business Leaders National, Inc.
busilead@kenbukan.org
270–723–7463

Cheri Lee
Texas Health Resources
cherilee@texashealh.org
817–462–7073

Johnny Lee
Peace at Work
jlee@peaceatwork.org
919–719–7203

Hank Linden
Longview Associates, Inc.
Hlinden@problemshavesolutions.com
914–946–0525

Herbert Linn
National Institute for Occupational Safety and Health
hlinn@cdc.gov
304–285–5947

Jane Lipscomb
University of Maryland
lipscomb@son.umaryland.edu

Rich Lombard
Unity Health System
rlombard@unityhealth.org
585–589–0662

Thomas Lowe
New York State Nurses Association
thomas.lowe@nysna.org
212–785–0157, Ext. 200

Wayne Lundstrom
WV Fatality Assessment and Control
 Evaluation (WVFACE)
wlundstrom@hsc.wvu.edu
304–293–1529

Jay Malcan
Virginia Union University
jmalcan@ci.richmond.va.us
804–646–6119

Daniel McDonald
Veterans Health Affairs
daniel.mcdonald@lrn.va.gov
205–731–1812

David McKay
Ohio Domestic Violence Network
pndmc@bright.net
937–492–9995

Kate McPhaul
University of Maryland Baltimore
mcphaul@son.umaryland.edu
410–706–4907

Dan Michael
Target Corporation
Dan.Michael@target.com
612–696–4133

Randall Miller
Baltimore County Police Department
rmiller@co.ba.md.us
410–931–2165

Sarah Miller
U.S. Department of Labor, Women's Bureau
miller.sarah@dol.gov
202–693–6716

Leah Morfin
Ms. Foundation for Women
lmorfin@ms.foundation.org
212–709–4405

Nancy Munro
American Association of Critical Care Nurses
jonamunr@hotmail.com
703–450–7911

Christine Neubauer
State Farm Insurance

Barry Nixon
National Institute for the Prevention of WPV,
 Inc.
wbnixon@aol.com
949–770–5264

Ellen Nolan
Prince William County Government
enolan@pwcgov.org
703–792–7967

Denise Null
General Motors
denise.p.null@gm.com
410–631–2103

John O'Brien
Veteran's Medical Center of Baltimore
John.Obrien@med.va.gov
410–605–7012

Emily O'Hagan
New Jersey Department of Health
emily.ohagan@doh.state.nj.us
609–292–9553

Anne O'Leary-Kelly
University of Arkansas
aokelly@walton.uark.edu
479–575–4566

Marc Oliver
University of Maryland
moliver@medicine.umaryland.edu

Richard Ottenstein
The Workplace Trauma Center
rjo@workplacetraumacenter.com
410–363–4432

Paul Papp
U.S. Army
Paul.Papp@usag.apg.army.mil
410–306–1079

George W. Pearson
TritonPCS/SunCom
gpearson@tritonpcs.com
804–364–7381

Corinne Peek–Asa
University of Iowa Injury Prevention Research Center
corinne-peek-asa@uiowa.edu
319–335–4895

Timothy Pizatella
National Institute for Occupational Safety and Health
tpizatella@cdc.gov
304–285–5894

Roderick Pullen
Community College of Baltimore County System
rpullen@ccbcmd.edu
410–455–4455

Susan Randolph
University of North Carolina School of Public Health
susan.randolph@unc.edu
919–966–0979

Deborah Reed
Illinois Nurses Association
debbireedrn@aol.com
217–523–0783

Carol Reeves
University of Arkansas
creeves@walton.uark.edu
479–575–6220

Chiara Rengo
Department Occupational Health Clinica Del Lavoro Luigi Devoto
omscons@unimi.it
0039/02/5454091

Joyce Renner
State Farm Insurance
joyce.renner.bh1q@statefarm.com
301–620–6130

William Rhoads
Centers for Disease Control and Prevention
wrhoads@cdc.gov
770–488–1284

Robyn Robbins
United Food and Commercial Workers Union
rrobbins@ufcw.org
202–466–1505

Roger Rosa
National Institute for Occupational Safety and Health
rrosa@cdc.gov
202–205–7856

Jonathan Rosen
New York State Public Employees Federation
Jrosen@pef.org
518–785–1900, Ext. 385

Benjamin Ross
Occupational Safety and Health Administration
rpss.benjamin@dol.gov
404–562–2284

Emily Rothman
Boston University School of Public Health,
   Department of Social and Behavioral
   Sciences
emfaith@aol.com
617–414–1385

Art Rudat
America Online, Inc.
arudat1@aol.com
703–265–5733

Eugene Rugala
earugala@fbiacademy.edu
703–632–4321

Robin Runge
American Bar Association Commission on
   Domestic Violence
runger@staff.abanet.org
202–662–8637

Georgia Sabatini
MBNA America
georgia.sabatini@mbna.com
410–229–6572

Vikki Sanders
OSHA Consultation, Minnesota
Vikki.sanders@state.mn.us
651–284–5274

Mario Scalora
University of Nebraska, Lincoln
mscalora1@unl.edu
402–472–3126

James Scaringi
Department of Veterans Affairs
james.scaringi@mail.va.gov
202–273–7381

Ronald Schouten
Massachusetts General Hospital/Harvard
   Medical School
rschouten@partners.org
617–726–5195

Mark Scovill
Texas Health Resources
MarkScovill@TexasHealth.org
817–462–7665

Rick Seta
New York Police Department
rick.seta@mbna.com
302–457–3242

Barbara Silverstein
Washington State Dept Labor and Industries
silb235@lni.wa.gov
360–902–5668

Rita Smith
National Coalition Against Domestic Violence
rsmith@ncadv.org
303–839–1852, Ext. 105

Kate Snyder
DAOHN
kathleen.snyder@mbna.com
302–432–0024

Rebecca Speer
Speer Associates
speer@workplacelaw.com
415–283–4888

Robert Stabler
Cape Canaveral Hospital
Bob.Stabler@Health-First.org
321–868–7235

Jennifer Stapleton
Corporate Alliance to End Partner Violence
jstapleton@domesticviolence.net
509–487–6783

Arnie Stenseth
Athena Research Corp.
spook163@athenaresearch.com
605–275–6028

Nancy Harvey
Steorts International
safety@crols.com
703–790–5116

Kiersten Stewart
Family Violence Prevention Fund
kiersten@endabuse.org
202–682–1212

Harley Stock
Incident Management Group
gbmi@aol.com
954–452–0434

Nancy Stout
National Institute for Occupational Safety and Health
nstout@cdc.gov
304–285–5894

Craig Swallow
craig.swallow@connexion2.com
+44 7968726891

Reena Tandon
Johns Hopkins School of Public Health
rtando@jhsph.edu

Linda Tapp
American Society of Safety Engineers
LTapp@crownsafety.com
856–489–6510

Robin Thompson
Robin H. Thompson & Associates
r-t@att.net

Corey Thompson
American Postal Workers Union, AFL-CIO
C_Thomspon@verizon.net
202–842–4273

Craig Thorne
University of Maryland School of Medicine
cthorne@medicine.umaryland.edu
410–706–7464

Phil Travers
Consumer Product Safety Commission
ptravers@cpsc.gov
303–504–7447

Glenn Valis
MBNA America
glenn.valis@mbna.com
410–229–6678

Dana Vogelsang
Florida Department of Health
Dana_Vogelsang@doh.state.fl.us
561–662–5647

KC Wagner
Cornell University–ILR
kcw8@cornell.edu
212–340–2826

Jane Walstedt
U.S. Department of Labor, Women's Bureau
Walstedt.Jane@dol.gov
202–693–6781

Dutchin Webster
CWA Local 2107
Dutchgirl418@aol.com
410–768–0611

Kim Wells
Corporate Alliance to End Partner Violence
kwells@caepv.org
309–664–0667

Deborah Widiss
Legal Momentum
dwidiss@legalmomentum.org
212–925–6635

Carol Wilkinson
IBM
drcarol@us.ibm.com
914–499–5555

William Zimmerman
United States Capitol Police Threat Assessment Section
william_zimmerman@cap-police.senate.gov
202–224–1495

www.ingramcontent.com/pod-product-compliance
Lightning Source LLC
Chambersburg PA
CBHW081753170526
45167CB00009B/4011